Alex Guimarães Sanches

# Agronomic performance of cassava genotypes in Baixo Amazonas, Pará

Alex Guimarães Sanches

# Agronomic performance of cassava genotypes in Baixo Amazonas, Pará

## Mandioculture in the Lower Amazon, Pará, Brazil

ScienciaScripts

**Imprint**



Cover image: www.ingimage.com

This book is a translation from the original published under ISBN 978-3-330-99574-1.

Publisher:
Sciencia Scripts
is a trademark of
Dodo Books Indian Ocean Ltd. and OmniScriptum S.R.L publishing group

120 High Road, East Finchley, London, N2 9ED, United Kingdom
Str. Armeneasca 28/1, office 1, Chisinau MD-2012, Republic of Moldova, Europe
Managing Directors: Ieva Konstantinova, Victoria Ursu
info@omniscriptum.com

Printed at: see last page
**ISBN: 978-620-8-58142-8**

**SUMMARY**


| | |
|---|---|
| **CHAPTER 1** | 3 |
| **CHAPTER 2** | 5 |
| **CHAPTER 3** | 12 |
| **CHAPTER 4** | 15 |
| **CHAPTER 5** | 18 |
| **CHAPTER 6** | 22 |
| **CHAPTER 7** | 25 |
| **CHAPTER 8** | 26 |
| **CHAPTER 9** | 30 |
| **CHAPTER 10** | 38 |

**ABSTRACT**: Introducing cassava cultivars into a given ecosystem and selecting the best adapted ones is a simple, low-cost procedure that should be carried out due to the high level of interaction between genotype and environment, as it is unlikely that a variety will behave in a similar way in all ecological regions. The aim of this study was to evaluate 30 varieties of cassava grown in the region known as the Lower Amazon, in the municipality of Monte Alegre-PA. The experimental design used was entirely randomised, with 30 treatments and three replications. The spacing used was 1.0 x 1.0 m, with 25 plants per plot. 1.0 $t.ha^{-1}$ of dolomitic limestone and 30 g of the formula 11:30:17 per plant were used. Average root yield was assessed using the Scott - Knott test at 5% probability, as well as agronomic characterisation in terms of the incidence of pests and diseases, plant size and the colour of the root pulp. Among the varieties with the highest root yields, Inajá, Chico vara, Inha, Saracura, Jurará amarela, Amolada, Kiriris and Bastiana stood out, producing more than 40.0 $t.ha^{-1}$ of roots. With regard to the stem borer, the most tolerant varieties were Saracura, Tomázia and Kiriris, while the most susceptible were Olho roxo, Santinha, Duquinha, Inajá and Pau velho. With regard to root rot, the varieties with the highest percentage of attack were Duquinha, Taxi vovó, Paulo velho and Jurará amarela.

Keywords: *Manihot esculenta.* Productivity. Root. Flour.

# CHAPTER 1

## INTRODUCTION

Pará is Brazil's main cassava producer. In 2013, 5,078,426 tonnes were produced in the state, in a planted area of 314,076 hectares, which corresponds to a productivity of 16.2 t/ha (IBGE, 2013).

In the state of Pará, cassava cultivation is mainly carried out by family farmers in order to guarantee their subsistence. Due to a lack of access to information and technologies, these farmers have been using itinerant land preparation practices over time, by slashing and burning, mainly primary forest and capoeiras. They also use the same varieties over time. The low level of technology applied contributes to low productivity.

Despite being one of the country's main producers, its productivity is very low in relation to the crop's potential, and production in the state is essentially used in the diet, in the form of flour.

Against this backdrop, the introduction of new varieties and the development of appropriate techniques for growing cassava under the soil and climatic conditions of each mesoregion are challenges for improving productivity and quality and turning cassava cultivation from a centuries-old subsistence crop into an excellent business, capable of meeting not only local demand, but also that of other regions, as well as generating a labour market and improving the quality of life of those involved in this activity.

The introduction and evaluation of new cassava varieties makes it possible to provide suitable varieties for each segment of the production chain, whether for making flour, starch, feed or other uses. This creates market and labour alternatives and strengthens the socio-economic development of the region.

SAGRILO et al. (2007) point out that introducing and evaluating new genotypes in new areas is the most common breeding method for selecting new cassava cultivars, as well as being the simplest and least expensive . Regional evaluation of new cultivars is necessary since, in the case of cassava, the interaction between genotype and environment is very pronounced. By using materials that have already been tested in other regions, the average time taken to obtain cultivars that meet farmers' specific needs is reduced.

According to Fukuda & Silva (2003), due to the high level of interaction between genotype and environment, it is unlikely that a variety will behave in a similar way in all ecological regions. Therefore, it is necessary to continually evaluate introduced varieties in comparison to local ones, in order to select those that best adapt to the soil and climate conditions of each region and provide better yields.

The aim of this work was to introduce and evaluate the root yields of 30 cassava varieties and define the best ones to be grown in the Baixo Amazonas Territory, Pará, Brazil.

# CHAPTER 2

## LITERATURE REVIEW

### 2.1 Botanical classification and ecophysiological aspects

Cassava is a dicotyledonous plant belonging to the Euphorbiaceae family and *the Manihot* genus. The genus comprises several species, of which *Manihot esculenta* Crantz stands out from an economic point of view, as it is highly hardy and adapts to different types of soil and climate (ALMEIDA et al., 1993).

It is an evergreen, monoecious, lactescent, erect shrub, reaching around 3m in height, with colouring that varies from green to reddish or violet tones. The root is fasciculate, tuberous, starchy and covered in a more or less thick tuber. The stem is brittle, with a thick pith and prominent scars left by falling leaves. The leaves are alternate, deciduous, long-petiolate, membranous, deeply divided into five to seven palmate segments, lanceolate, acuminate, with a smooth edge and deciduous stipules. The male flowers are smaller and more numerous and the female flowers, two or more in number, are found at the base of the inflorescence (ALMEIDA et al., 1993).

Cassava originated in South America, mainly in the Amazon region, with Brazil as its country of origin. However, Africa is by far the continent with the largest production, which shows that this crop is widespread in many regions of the planet, as it is very tolerant of adverse climate and soil conditions. It is grown in regions with a tropical and subtropical climate, with annual rainfall varying between 600 and 1,200 mm (CEPLAC, 2007), as well as in regions with an equatorial climate, with annual rainfall of over 2,500 mm, which is the case in the Brazilian Amazon.

Seed propagation is only used for experimental and breeding purposes. For commercial crops, the vegetative propagule is the stem of the plant itself, obtained from the middle third of the shoot. The stem is cut into small segments 15 to 20 cm long with five to **seven buds, called -manivall (MATOS & CARDOSO, 2003).**

Cassava belongs to the group of cyanogenic plants because it has cyanic compounds and enzymes distributed in varying concentrations in the different parts of the plant. By breaking down the cellular structure of the root, the enzymes present (linamarase) degrade these compounds, releasing hydrocyanic acid (HCN), which is the toxic principle of cassava and whose ingestion or even inhalation represents a serious health hazard, with extreme cases of poisoning occurring. The lethal dose is considered to be approximately 10 mg of HCN per kg of live weight (CAGNON et al., 2002).

According to Cagnon et al. (2002), in relation to the hydrocyanic acid content in the root, cultivars are classified as tame, when they have less than 50 mg of HCN/kg of fresh root without the peel; moderately poisonous, when they have between 50 and 100 mg of HCN/kg of fresh root without the peel; and wild or poisonous, above 100 mg HCN/kg of fresh root without the peel, with tame cultivars also being known as table, cassava and yuca (CAGNON et al., 2002).

According to Mattos & Cardoso (2003), HCN is found in all parts of the plant. The leaves have the highest percentages. In the root, the cortex or thick bark contains higher levels than the pulp, and the cortex **of the roots of the -mansall variety** may **contain a higher percentage than the pulp of the roots** of the "bravas" varieties. Cassava varieties with high HCN levels can be used for industrial purposes because the HCN content in the roots is released during processing, but cooking may not eliminate all the acid, which is why "wild" cassava, even if cooked, can cause accidents, which is not the case with industrial products that are subjected to higher temperatures and for longer. Drying (by the heat of the sun or dryers) eliminates the acid by volatilisation.

### 2.2 Economic importance

Cassava production is located in tropical areas, which is why it is absent from the list of plants grown in most industrialised countries. The five main producing countries are, in order of importance: Nigeria, Brazil, Thailand, the Democratic Republic of Congo and Indonesia (FAOSTAT, 2008).

According to Treche (1995), cassava is considered the fourth most important crop in the world, after rice, wheat and maize, due to its contribution to human nutrition, as well as its great added economic value due to the various ways in which it can be utilised. Everything can be utilised from cassava, but consumption of the roots worldwide is much more significant, with the main by-products being starch and flour.

Compared to other crops, cassava has a number of advantages, such as: easy propagation, high tolerance to long droughts, satisfactory yields even in low-fertility soils, low demands on agricultural inputs, potential resistance or tolerance to pests and diseases, high starch content in the roots and the possibility of using it from root to leaf. Cassava is used in the production of human food products and animal feed, using the root and the aerial part of the plant (branches and leaves) **to be fed to animals -in naturall, in the form of slices, crushed or dehydrated bran, as a component of** feed in pig, poultry and cattle farming (CEREDA, 2005).

There has recently been an improvement in the sensory quality of the flour produced by large industries and an increase in the number of applications for cassava or its products, such as starch, shavings, dextrins, glues, the paper, cardboard and fabric industries, the production of glucose, alcohol and acetone, among others. With the evolution of eating habits, industrial needs and market requirements, the opening up and search for markets, new uses are emerging (FOLEGATTI **et al.**, 2005).

Although Brazil is the largest producer in Latin America, the crop's yield is not the highest on the continent, and Brazilian productivity is lower than in Paraguay. According to Cock (1979), manioc can have a production potential of 90 tonnes/ha, but the Brazilian average is only 14 tonnes/ha (FAO, 2007).

When it comes to the distribution of production among Brazil's different regions, the Northeast and North stand out from the rest, with a 46% and 25% share respectively. The South has a 17% share, the Southeast 7% and the Centre-West 4% (IBRE/FGV, 2005 cited by FURLANETO, 2006). In the South and Southeast, most production is orientated towards industry, mainly in the states of Paraná, São Paulo and Minas Gerais.

According to DINIZ et al. (2004), cassava farming provides annual gross revenue equivalent to 2.5 billion dollars to the world economy and a tax contribution of 150 million reais to Brazil. Production for the manufacture of flour and starch generates gross annual revenue equivalent to 600 million and 150 million reais respectively, demonstrating that the crop makes a significant contribution to Brazil's

Gross Domestic Product (GDP).

In 2000 and 2001, Brazil exported an average of 13.1 million tonnes of cassava starch. A total of 79.9% of exports went to Venezuela (31.4%), Argentina (26.8%), Colombia (10.7%), Uruguay (6.2%) and the United States (4.8%). Of all the starch exported, 6.2 per cent comes from the Cerrado region. In the form of fresh, chilled, frozen or dried cassava root, Brazil exported an average of 361 tonnes during this period. The countries that buy the most
79.5% of exports are: United Kingdom (40.2%), United States (14.0%), France (13.0%) and Japan (12.3%). Of all the roots exported, 2.4% came from the Cerrado region: the states of Bahia and Goiás (ABAM, 1998).

### 2.3 Nutritional importance

According to Mattos and Cardoso (2003) the cassava plant can be used in its entirety to feed various species of domestic animals such as cattle, buffalo, poultry and pigs. The roots are a source of carbohydrates and the aerial part, as well as carbohydrates, also provides protein, especially in the leaves. For animal feed, the hydrocyanic acid content must be low, both in the leaves and the roots, to avoid poisoning the animals. When wild varieties are used in animal feed, they should be given in the form of dehydrated shavings or silage.

According to Sousa and Fialho (2007) the roots are an important food source, rich in calories and carbohydrates, and can be used for human and animal consumption, as well as being processed in various ways, such as: cooked root, flour, tucupi, starch, glues, paper, cardboard, fabrics, alcohol production and acetone, etc.

However, according to CORREA et al. (2005), the cassava root can provide around 1,460 Kcal/kg of fresh weight and can be considered, in nutritional terms, as a cheap source of energy, requiring other foods as a source of protein, vitamins, minerals and fat to meet human nutritional needs. The leaves contain a significant amount of vitamins.

CEREDA (2001) reports that cassava has a low vitamin content, although yellow varieties can have appreciable levels of carotene, as well as vitamin C, which is largely lost during processing. However, the consumption of cassava or its derivatives should always be accompanied by protein-rich foods, such as meat and fish, as it is essentially an energy food with a low protein content.

The cassava plant is known to have a number of anti-nutritional factors such as tannins, saponins, oxalic acid, nitrates, lectins, phytates, among others, and the toxic one, hydrocyanic acid (HCN) (CARVALHO, 1978).

HCN is present in all parts of the plant and varies according to cultivar, age and location on the plant, available water, nitrogen fertilisation and shading (CARVALHO, 1978). It is highly volatile and disperses easily when the product is handled, whether through drying (CURCELLI et al., 2009; VALDIVIÉ et al., 2011), grinding or fermentation processes. This is because the cassava plant contains cyanogenic glycosides, particularly Linamarin and Lotaustralin. These receive the utmost attention because when hydrolysed by the enzymes Linamarase and Lotaustralase, which are present throughout the plant, they release hydrocyanic acid (HCN) or free cyanide, which can intoxicate and even lead to death when they exceed tolerable intake limits (VALDIVIÉ et al., 2011).

Intoxication of homeothermic animals occurs because HCN reacts with the ferric form of the iron ion in haemoglobin, making it impossible to transport oxygen and causing asphyxiation. HCN also forms a complex with copper-containing enzymes, limiting important intermediate metabolism reactions (VALDIVIÉ et al., 2011).

Wanapat (2009) reports that hay from the leaves and upper branches of cassava plants harvested 3-4 months after planting can contain approximately 3.9% condensed tannins. According to the same author, tannin levels tend to be higher in older cassava leaves and lower in young leaves. However, cassava foliage, meal and hay have appropriate tannin content for rabbits, 3.8 to 5.5 %MS, which can generate high substitution rates for grass hay (which generally raises the price of feed), but high for pigs, poultry and other non-herbivorous monogastrics, leading to a limited inclusion of this ingredient in rations for these animals.

**2.4 Social importance**

Cassava is grown all over Brazil and represents income and food for farmers, regardless of soil type, fertility and climatic variations. According to GIRAUD et al. (1995), cassava is produced in more than 100 countries and consumed by thousands of people around the world, mainly on the African and Asian continents and in South America. It is a staple food for the poor populations of these continents and plays a fundamental role in the fight against the nutritional problems that plague them.

**The "casas-de-farinha" or "farinheirasll" are generally simple buildings of** great social importance to the small producers who live around them, as **these "farinheirasll" are directly or indirectly used by family members and neighbours from the area where** they are installed (located in the production units), becoming a place of sociability, men and women, adults, young people and children take part in planting, harvesting, making flour, tucupi and tapioca gum, and are often joined by neighbours from the community, often bringing together work and leisure activities (CARDOSO, 2005).

There are three basic situations in the cassava production system and marketing chain, which take into account the interconnections between the origin of the labour force, the level of technology, market share and the degree of intensity of the use of capital on the farm. These are: the domestic unit, the family unit and the business unit. Mattos & Cardoso (2003) define them as: the domestic unit is characterised by the use of family labour, does not use modern technology, participates little in the market and has a low level of working capital; the family unit, unlike the domestic unit, adopts modern technology, participates significantly in the market and has a higher level of working capital; the business unit is characterised by the hiring of third party labour, the adoption of modern technology and has a high level of participation in the market.

According to Cardoso (2005), casas-de-farinha are part of the regional rural scene, where their presence is strongly associated with colonisation in the Amazon, as they represent the population's food base, a situation made possible by the plant's ease of cultivation, producing in adverse conditions, as well as finding favourable environmental conditions for its cultivation and development in the northern region.

There is a great disparity in the per capita consumption of cassava flour between

Brazilian regions. The Brazilian average is 7.8 kilos per capita per year, which is "inflated" by the 33.8 kilos per capita in the North. The Northeast consumes 15.3 kilos per capita per year, while the other regions consume less than 1.5 kilos per capita per year. The state that stands out in this index is Amazonas, with 43.4 kg/capita/year, in contrast to Paraná's 0.7 kg/capita/year, despite its considerable production (IBGE, 2010).

## CHAPTER 3

### CULTIVATION SYSTEM

The first step in setting up manioc farming, as with any other crop, is preparing the area. In family farming, this is done in the traditional slash-and-burn way in areas of primary forest or capoeira. It is usually a succession crop where cassava is the second crop to be planted in the area, usually preceded by annual crops (TONTINI, 2009; SANTOS, 2007).

The slash-and-burn system in the practice of growing food crops has drawbacks such as environmental pollution (through burning), erosion, loss of nutrients, as well as being dangerous and arduous work, with great physical wear and tear on the farmer (MATTOS and CARDOSO, 2003).

When preparing the soil, a soil analysis should be carried out to guide acidity correction and fertilisation in accordance with the recommendations for cassava cultivation, in order to allow the plants to grow quickly and cover the soil more quickly. In the state of Pará, planting usually takes place in two different climatic periods: at the beginning of the rainy season, in December or early January, and the **summer planting at the end of the rainy season, in May and June, when the** rains decrease in intensity and frequency. If there is a risk of excess moisture in the soil, planting can be carried out after the rains have started or the manioc can be planted vertically, leaving around 5 cm under the soil (MATTOS and CARDOSO, 2003).

According to Câmara et al. (1994), selecting branches and obtaining good quality maniocs are management mechanisms capable of significantly increasing the yield of cassava roots. According to Oka et al. (1987), when selecting cuttings, the density must be taken into account, i.e. the ratio between the weight and volume of the cuttings, characteristics that determine their ability to sprout. In addition to these aspects, the age of the plant, the diameter of the stem and the part of the plant to be used as planting material are of great importance when choosing the branches.

Healthy plants between 8 and 18 months old, with mature and well developed branches, provide good quality planting material (CORRÊA & VIEIRA Neto, 1978). Younger plants, due to their high proportion of green and tender tissues, and older plants, due to their intense lignification, are not ideal for supplying cuttings for planting. Although they do sprout, cuttings from green and not very lignified plants are very susceptible to attack by pathogens and insects and dehydrate very quickly. On the other hand, plants older than 18 months have very lignified lower thirds, which results in late sprouting cuttings and less vigorous shoots (LOZANO et al., 1982 ).

Along with selection and preparation, the length of the maniocs and the number of nodes are of great importance for the proper establishment of the crop. Maniples that are 20 cm long, with at least five nodes, generally have a higher sprouting rate, allowing for the development of more vigorous plants, which results in a higher yield of roots (MATTOS, 1993; SOUZA, 1993), although this length can vary according to the planting position and the variety (CHAN et al., 1993).

MOHD,1985). When planted horizontally, maniocs that are 20 cm long result in greater tuberous root production and, for inclined planting, maniocs that are 30 cm long produce more (VIANA et al., 2000).

One of the factors that determines the quality of cassava planting material is the diameter of the stem, which is linked to its age. A practical observation to assess the degree of maturity is to check whether the diameter of the pith is equal to or less than 50 per cent of the diameter of the stem, in cross-section. If so, the stem is at the appropriate stage of maturity to be planted (CONCEIÇÃO, 1983). As a general rule, it is recommended that the diameter of the selected cutting should not be less than half the diameter of the thickest part of the stem of the variety being used (LOZANO et al., 1977).

When the reserve roots are filled, carbohydrates are directed from the leaves to the roots. Starch deposition begins 25 days after planting (CONCEIÇÃO, 1983). Cassava roots have an average composition of 68.2% moisture, 30% starch, 2% ash, 1.3% protein, 0.2% lipids and 0.3% fibre (PADONOU et al., 2005).

For human consumption, the main characteristic is that the cultivars have less than 100 ppm or 100 mg of hydrocyanic acid (HCN) per kilogram of raw root pulp.

The HCN content varies between cultivars, with the environment and with the physiological state of the plant, and is a decisive factor when choosing a cassava cultivar (PADONOU et al., 2005).

# CHAPTER 4

## ROOT CHARACTERISTICS

### 4.1 Humidity

Determining moisture in food is of great importance because water has a marked influence on several of its characteristics, such as the appearance, flavour, structure, susceptibility and deterioration of food. Water solubilises important compounds such as vitamins, minerals, sugars and acids. Therefore, knowledge of the moisture content of food is extremely important (BOBBIO and BOBBIO, 1992). Food preservation can depend on the moisture content present in the material.

Furthermore, when comparing the nutritional value of two or more foods, the respective dry matter contents must be taken into account. 25 Usually, the water content of a food is expressed by the value obtained when determining the total water contained in it. However, this value does not give any indication of how the water is distributed, nor does it allow us to know whether all the water is bound in the same way to the food (BOBBIO and BOBBIO, 1992).

There are at least two types of water in food: one is called free water, which is weakly bound to the substrate and acts as a solvent, allowing microorganisms to grow and chemical reactions to take place, and which is eliminated relatively easily, and the other is combined water, which is strongly bound to the substrate, more difficult to eliminate and which is not used as a solvent and therefore does not allow microorganisms to grow and slows down chemical reactions (BOBBIO and BOBBIO, 1992).

The loss of water from cassava roots during storage has a negative impact on the initial development of vascular discolouration (MARRIOT; BEEN; PERKUINS, 1978). Therefore, humidity affects deterioration (RICKARD, 1984). When roots are stored under low relative humidity, a dry white layer, a few millimetres thick, develops during the first 72 hours, and vascular striation or discolouration is also observed along the roots (RICKARD, 1985). At high relative humidities, there is condensation of water on the surface of the product and infestation by a large number of pathogens, resulting in tissue death.

### 4.2 Cooking time

Cassava cultivars vary in cooking time (BORGES et al., 1992) and it has been shown in 26 studies that this factor varies with the type of soil, the cultivar and the age of the plant (FUKUDA and BORGES, 1990). Cooking time is a fundamental characteristic for selecting a cultivar, and the one with the shortest cooking time is preferred (FUKUDA and BORGES, 1990).

The variability and instability of the cooking time of cassava roots is not well known (LORENZI, 1994). The variables related to cooking that have been most important in the research sector are the texture, plasticity and stickiness of the cooked dough, as they directly interfere in most culinary recipes prepared with cassava (PEREIRA; LORENZI; VALLE, 1985). However, Lorenzi (1994) states that these variables are associated with the length of the cooking time and the shorter this time, the better the sensory qualities of the pasta produced. The use of the Mattson cooker in the methodology for determining cooking time provided a more objective assessment than that proposed by Pereira, Lorenzi and Valle, (1985), Carvalho et al. (2000) and Lorenzi et al. (1996), in which cooking was assessed subjectively with the aid of a fork.

The cooking time for roots of good culinary quality should not exceed 30 minutes and the cooked pulp should be easily crushed and broken up when mashed with a fork, forming a paste with a mealy texture and a plastic, mouldable consistency (BORGES; CARVALHO; FUKUDA, 1992).

### 4.3 Starch

The starch stored and extracted from cassava roots is classified as a polysaccharide. It provides between 70 and 80 per cent of the calories consumed by humans. The most important potential sources of starch are cereal grains (40% to 90% of their dry weight), legumes (30% to 70% of their dry weight) and tubers (65% to 85% of their dry weight) (FREITAS et al., 2003). They are made up of linear and branched macromolecules (CEREDA et al., 2001), known as amylose and amylopectin.

Amylose is a linear molecule made up of D-glucose units linked **at a,** 1-4. Amylopectin, on the other hand, is highly branched and made up of D-glucose units **linked at a,** 1-4 **and with 5% to 6% of a,** 1-6 **links** at the branch points. It is also called starch when dried in dryers, sweet starch or gum when dried in the sun, and is obtained by processing cassava roots that have been properly cleaned, peeled, crushed, disintegrated, purified, sieved, centrifuged, concentrated, dehydrated and dried (BOURSIER, 1994). It is an extremely versatile product and is widely used in the food, textile and pharmaceutical industries (ARIENT et al., 2005).

Due to its low cost, starch is widely used by the food industry as a calorific ingredient and as an enhancer of physicochemical properties. Various characteristics, such as texture, appearance, moisture, consistency and storage stability (shelf life), are altered and even controlled with the use of starch. It can also be used to bind or disintegrate, expand or thicken, lighten or make opaque, retain moisture, produce a smooth or pulpy texture and light or crunchy toppings. It is also used both to stabilise emulsions and to form oil-resistant films (CEREDA et al., 2001).

# CHAPTER 5

## MAIN PESTS THAT ATTACK CASSAVA

### 5.1 Lace bug

The species commonly known as lace bugs are reported to be associated with cassava cultivation in several Latin American countries (BELLOTTI, 2002). It belongs to the family Tingidae (Hemiptera. Heteroptera), genus *Vatiga*, with five species described for the Neotropical region, *V. illudens, V. manihotae, V. pauxilla, V. variante and V. cassiae* (FROESCHNER, 1993). Bellotti et al. (1999) cite the predominance of V. *illudens* in Brazil, although according to the author the species *V. manihotae, V. variante and V. cassiae* also occur.

The adult insects of the lace bug are grey in colour, while the nymphs are whitish (LOZANO et al., 1983) with the antennae and encircles darkening as they develop.

These insects are initially located on the underside of the basal and middle leaves of the plant, but in high populations they reach the apical leaves and are favoured by periods of drought (BELLOTTI et al. 2002b). They feed on the protoplast of leaf prenchyma cells and consequently initially leave chlorotic spots on the leaves, which can develop into reddish-brown colours (BELLOTTI, et al., 1999). Due to the lesions caused by the pest, photosynthesis is reduced, the lower leaves fall off and, in the case of severe infestations, the plant can be completely defoliated (BELLOTTI et al., 2002c).

When it comes to assessing yield reduction, the results vary, with some studies pointing to a reduction of 55% (FIALHO et al., 2009), 39% (BELLOTTI et al., 1999), 21% (FIALHO et al., 1994) and no reduction in yield (MARTINAZZO et al., 2007b). As for damage to the aerial part, studies indicate between 48 and 50 per cent reduction in leaves in the upper third of the plant (FIALHO et al., 2009).

Studies indicate that the damage caused by this insect can be influenced by the levels of cyanogenic compounds in the plant. Work carried out with tame and wild cultivars showed a reduction in the level of infestation in varieties with hydrocyanic acid levels above 100 ppm (OLIVEIRA et al., 2005; SANTOS et al.,2008a ; FIALHO et al.,2009).

The best recommendation at the moment is to adopt good management practices that contribute to the good development of the plants, such as soil correction and fertilisation, obtaining branches for planting in non-infested areas, selecting mani-seeds, planting in multiple or intercropped crops and crop rotation (FIALHO et al., 2001).

**5.2 Mealybugs**

According to BELLOTTI et al. (2002b) there are more than 15 species of mealybugs feeding on cassava distributed throughout Africa and Latin America, but only the species *Phenacoccus herreni* and P. *manihoti* are of economic importance, both being of tropical origin.

These species have similar cycles, differing only in the fact that P. *herreni* reproduces sexually, with the presence of males, while P. *manihoti* reproduces via parthenogenesis, i.e. females giving rise to females (BELLOTTI et al., 2002b).

Damage is caused by both the young and adult stages of the mealybug and is characterised by direct damage due to sap suction, leaving the plant weakened and looking nutritionally deficient; damage due to saliva toxicity, caused mainly in the young regions of the plant, causing deformation of the shoots which become shrivelled with a cabbage-like appearance and causing the leaves to curl and fall prematurely and, in high populations, causing necrosis of the apical tissues and consequent death of the tops (BELLOTTI et al, 1999, 2002; BENTO et al., 2002).

As for its damage potential, an average reduction of 76 per cent was observed in the weight of the roots of plants from maniocs infested with the mealybug when compared to those from healthy maniocs in the Brazilian cerrado region (OLIVEIRA et al. 2005; OLIVEIRA E FILHO, 2005b).

Control of these species is carried out by natural enemies such as polyphages and oligophages, which prey on mealybugs (NEUENSCHWANDER, 2001). Among these, ladybirds are the most important regulators when high mealybug populations occur, but they are less efficient when the pest population is low (HERREN AND NEUENSCHWANDER, 1991).

Considering that one of the main ways of spreading the pest is through contaminated vegetative material, one of the main means of control is to select branches with good sanitary conditions that do not come from regions where the pest is present (BELLOTTI et al, 1999). The authors also suggest destroying the plants in the initial outbreaks of the mealybug, in which case it is important to carry out a detailed inspection of the areas with a history of attack. In addition to the cultural practices mentioned above, proper soil correction and fertilisation is important to reduce the impact of the pest. Poor soils favour the mealybug population and disfavour the parasitoid A. *Lupezi* (BELLOTTI et al, 1999).

### 5.3 Stem borer

Stem borers are found in practically all cassava-producing regions of the world, but are particularly important in the Americas, especially in Brazil. They generally cause sporadic or localised damage. The larvae vary in size and shape according to the species and are found tunnelling in the centre or pith of the stems, which can then dry out and break by the wind. Excrement and exudates that fall from the galleries excavated by the larvae can be found on the infested stems or on the ground next to the plants. The adults feed on the apexes of the young stems, which slows down the plant's growth (NEUENSCHWANDER, 2001). During dry periods, attacked plants can lose their leaves and dry out. When the infestation is severe, the plants may die.

In Brazil, there are no records of a reduction in root yield, but there is a reduction in the quantity and quality of propagation material. In a study carried out in Cruz das Almas, BA, at Embrapa Cassava and Fruit Growing, during a period in which there was a severe attack by the stem borer, the Cassava Germplasm Bank was evaluated and 20 varieties were selected among the most attacked. The seed maniocs came from attacked stems and were planted without fertiliser. It was found that there was a reduction of up to 41 per cent in sprouting, varying between the cultivars. As the assessments were made, the percentage of plant mortality increased. Thus, there were reductions of 47%, 50% and 57% at 30, 60 and 130 days, respectively, after sprouting. It was observed that the plants that survived did not show normal development (stunted), affecting the quality of the stems and, consequently, the yield of roots (EMBRAPA, 1999).

It is recommended to periodically check the manioc, especially during the summer. Attacked stems should be cut and burnt in order to prevent the larvae from developing, which reduces pest populations. It is also recommended to use healthy maniocs for planting, which should be done through careful selection. Always try to use material from plantations where the stem borer has not been attacked (EMBRAPA, 1999).

#### 5.4 Cassava Mandarin

The caterpillars that feed on Euphorbiaceae, *Erinnyis ello*, are known in **Brazil by the common names of -gervãoll and -mandarovâll; they are also called -marandovâll. Such a serious pest, which causes so much damage to farmers, requires** serious care to combat (FERREIRA, 1941).

The gervão eats the leaves of cassava and yucca, as well as the tender parts of the stalks. When it attacks new plantations that are less than a year old, it often causes their total death. In older plantations, it slows down growth and greatly reduces root production (FERREIRA 1941).

FERREIRA LIMA, 1941, pointing out the damage caused by the caterpillar in manioc plantations in Santa Catarina and discussing the means of combating it, reports that according to calculations made by local farmers, the damage caused by the caterpillar is frightening. Thus, a normal field that would produce an average of 300 kilos of flour, when attacked by the caterpillar, produces an average of 90 kilos, in other words, **an average of 70% of the production of a cassava field is lost.**

## CHAPTER 6

### MAIN DISEASES THAT ATTACK CASSAVA

#### 6.1 Bacteriosis

Bacteriosis, caused by *Xanthomonas axoponodis* pv. manihotis, is the main disease of cassava, especially in the South, Southeast and Centre-West. The symptoms of bacteriosis are characterised by angular, watery spots on the leaves, wilting of the leaves and petioles, downward death and gum exudation on the stems, as well as necrosis of the vascular bundles and death of the plant. The damage caused by bacteriosis varies according to climatic conditions, the susceptibility or tolerance of the varieties, the cultural practices employed, the planting times and the level of contamination of the planting material (ABAM, 1998).

The abrupt variation in temperature between day and night is the most important factor for the severe manifestation of the disease, with a daily temperature range of more than 10°C for a period of more than five days being the ideal condition for the disease to fully develop. Production losses are around 30% in crops using susceptible varieties and, in places with favourable conditions for the disease, losses can be total. On the other hand, using tolerant varieties, even with favourable conditions, production losses reach a maximum of 30% (EMBRAPA, 2003).

. The use of resistant varieties is the most efficient measure for controlling bacteriosis, but cultural practices such as the use of healthy maniocs and appropriate planting times also contribute (EMBRAPA, 2003).

The varieties currently in use in areas where bacteriosis occurs are characterised by acceptable tolerance to the disease. In the Centre-West Region (Mato Grosso do Sul, Mato Grosso, Goiás, Distrito Federal and Tocantins), the recommended varieties are IAC-12, IAC-13, IAC-14, Mantiqueira, Branca de Santa Catarina, IAC 352-6, IAC 352-7, IAC 12-829, IAC 7-127, Sonora, EAB 81 and EAB 653 (EMBRAPA, 2003).

#### 6.2 Root rot

Root rot, caused by the fungi *Phytophthora* sp. and *Fusarium* sp., is a disease of little importance in the Cerrado, being more limiting in the Northeast, where losses in root productivity in the areas where it occurs most are around 30 per cent. In the North, it is particularly important in the Várzea and Terra Firme ecosystems, where losses can exceed 50% in the former and reach up to 30% in the latter. In some cases, total losses have been observed, especially in areas with dense soils and subject to constant waterlogging (EMBRAPA, 2003).

The symptoms of root rot are quite different depending on the causal agents. Typically, *Phytophthora* sp. attacks the crop in its adult stage, causing soft rot on the roots, with a very strong odour, similar to that of decaying organic matter; they show a greyish colour that is made up of the mycelia or even spores of the fungus in the affected tissues. The appearance of visible symptoms is more frequent on mature roots; however, there are cases of symptoms appearing at the base of young stems or on newly sprouted plants, causing wilting and total death. In the case of *Fusarium* sp., symptoms can occur at any stage of the plant's development and rarely cause direct damage to the roots (EMBRAPA, 2003).

The attack occurs at the point of the stem close to the ground, causing infections and often completely obstructing the vascular tissues, preventing the free circulation of sap and consequently causing indirect root rot. Unlike *Phytophthora* sp., the symptoms caused by *Fusarium* sp. attack on the roots are characterised by dry rot and no apparent disturbance of the tissues (EMBRAPA, 2003).

Root rot control measures involve integrating the use of tolerant varieties, associated with cultural practices such as crop rotation, physical and chemical soil management, cultivation systems and others. In the northern region, research carried out in the floodplains has shown that the use of a tolerant variety, combined with crop rotation and planting systems, has reduced root rot by around 60 per cent. The varieties considered tolerant so far are: Osso Duro, Cedinha,
Bibiana, clone 148/02, Aramaris and Kiriris, for the Northeast, and Zolhudinha, Mãe Joana and Embrapa 8, for the North (EMBRAPA, 2003).

### 6.3 Super Stretch

Super elongation, caused by Sphaceloma manihoticola, is one of the most important fungal diseases of the cassava crop. Its occurrence in Brazil was first noted in 1977 in the northern region, in Amazonas and Pará. After almost ten years it was detected in Mato Grosso, attacking large commercial crops. The main symptoms are characterised by the exaggerated elongation of tender or developing stems, forming thin branches with long internodes (EMBRAPA, 2003).

In severe cases, affected plants can be identified by the typical warty lesions on the stems, petioles and veins; leaf twisting, defoliation and tissue death are also common. The disease spreads very quickly during the rainy season, as the spores are easily transported long distances by wind and rain. The establishment of the disease in areas free of it occurs mainly through contaminated seed maniocs. The control measures for over-extension are basically the selection of healthy maniocs for planting, the elimination of infected plants, the use of tolerant or resistant cultivars and crop rotation in previously affected areas (EMBRAPA, 2003).

## CHAPTER 7

### VARIETY EVALUATION

With the aim of assessing the root yield and flour yield of 15 regional cassava varieties, Guimarães et al. (2009) carried out a study in the experimental area of the State University of Southwest Bahia, in Vitória da Conquista. In order to simulate normal growing conditions in the region, the soil was ploughed and harrowed, but without liming or fertilisation, and they used a spacing of 1.0 x 0.6m. This research found root yields ranging from 5.7 to 18.7 $t.ha^{-1}$, flour yields from 18.9 to 25.0% and flour productivity ranging from 1.2 to 3.9 t.ha-i. They also found that of the most productive varieties in t.ha -i of roots, only Platinão remained among those with the highest flour yields (%). Even so, those with the highest root yields were the ones that provided the highest flour yields in t.ha-i. On the other hand, the varieties with the lowest root yields (t.ha-i) were also those with the lowest flour yields (t.ha-i). The Platinão variety stands out in this research, as it was significantly superior in all three aspects assessed.

Rodrigues et al. (2007) evaluated the root yields of 10 cassava varieties in a sandy-textured, dystrophic yellow latosol in the municipality of Salvaterra, PA. The area was prepared with a plough harrow. A spacing of 1.0 x 1.0 m was used, with the application of 2.0 $kg.cova^{1}$ of organic compost when planting and top dressing with the equivalent of 600 kg.hai of NPK, formulation 10:28:20, 30 days later. Root yields ranged from 17.2 to 54.1 t.ha-i with the Pai Mane and Roxinha varieties, respectively.

# CHAPTER 8

## MATERIAL AND METHODS

### 8.1 Location and characterisation of the study area

To conduct the experiment, an experimental area was set up on the campus of the Technological School of the State of Pará, EETEPA, located in the municipality of Monte Alegre-PA.

Silva et al. (2015) evaluated meteorological data from the city of Monte Alegre, PA, for the period 1990 to 2002. The results indicated that the average annual rainfall is 2,123 mm, characterised by a rainy period between January and May, with 74% of the rainfall, March being the month with the highest rainfall, with an average of 379.2 mm. Only 14% of the rain falls between July and November, with August being the driest month with an average of 22.5 mm. The annual average daily temperature for the period was 27.3°C, with average maximum and minimum temperatures of 32.4°C and 22.1°C, respectively. The average monthly relative humidity was 80.4 per cent. The month of July is the period of greatest insolation, with 7.4 hours of average daily sunshine.

The crop was planted on a sandy-textured Yellow Dystrophic Latosol, which, according to EMBRAPA's National Soil Research Centre (1999), comprises mineral soils at an advanced stage of weathering, with a low cation exchange capacity, and are generally strongly acidic, with low base saturation.

### 8.2 Origin of the varieties and preparation of the manioc seeds

A total of 30 cassava varieties were evaluated, 25 of which were supplied by EMBRAPA Amazônia Oriental, from the Germplasm Bank located in the city of Belém, PA, and five (05) obtained from farmers neighbouring the experiment site, as shown in Table 01. The maniocs were obtained from healthy plants aged between 12 and 16 months. Only the maniocs located in the middle third of the plants were selected for the experiment (Figure 2.a attached), cut into sizes equal to 20 cm and used those with a diameter of approximately 2.0 cm, with five to seven vegetative buds (CORREA and ROCHA, 1979).

### 8.3 Setting up the experiment

The experiment was set up in a completely randomised design with 30 treatments (varieties) and three replications, totalling 90 experimental plots. Each plot was established in a 5 m x 5 m area, with the holes spaced 1.0 m x 1.0 m apart, totalling 25 plants per plot. The useful plot used to obtain the average root yield was made up of the nine (09) centre plants of the total plot. The other plants on the periphery of each experimental plot were used to act as a border and mitigate external effects.

The experimental plots (25 $m^2$) were separated from each other by a strip of 2.0 m longitudinally and 3.0 m transversely. The total area used for the research was 107 m x 49 m. The treatments applied with the identification of the respective cassava varieties are listed in Table 01.

The experimental area was prepared on 19/04/2009 using Rhoundup WG at a dose of 5.0 mL per litre of water to desiccate the Brachiarão. After desiccation, harrowing was carried out on 20/05/2009 to incorporate the plant remains and level the soil. On the same day, the experimental plots were marked out and 1.43 $t.ha^{-1}$ of dolomitic limestone with a PRNT of 70% was applied, equivalent to 1.0 $t.ha^{-1}$ with a PRNT of 100% (figure 2.b attached). Planting took place on 21 May 2009. Thirty days after planting, fertiliser was applied with 30 g per plant of the 11:30:17 formulation.

The area was cleaned with two manual weeding operations and two manual weeding operations in the third and sixth months and in the ninth and twelfth months after germination, respectively. Pest control was carried out systematically, reporting attacks by leaf-cutting ants using formicide bait and mandarová using Lorsbam at a dose of 2.5 mL per litre of water. There were no significant attacks of diseases of the aerial part that warranted control. The varieties were harvested 12 months after germination. They were then identified and weighed.

**Table 1:** Identification of the treatments and varieties used in the experiment.

| TREATMENTS | VARIETIES | ORIGIN | TREATMENTS | VARIETIES | ORIGIN |
|---|---|---|---|---|---|
| TRAT-01 | Ceará | EMBRAP | TRAT-16 | Honed | EMBRAP |

| | | A | | | A |
|---|---|---|---|---|---|
| TRAT-02 | Saracura | EMBRAPA | TRAT-17 | BRS Mari | EMBRAPA |
| TRAT-03 | Pacajás | EMBRAPA | TRAT-18 | Kiriris | EMBRAPA |
| TRAT-04 | Green Eye | EMBRAPA | TRAT-19 | BRS Poti | EMBRAPA |
| TRAT-05 | Jurará | EMBRAPA | TRAT-20 | Paulo Velho | EMBRAPA |
| TRAT-06 | Inha | EMBRAPA | TRAT-21 | Ouro Preto | EMBRAPA |
| TRAT-07 | Chico Vara | EMBRAPA | TRAT-22 | Maraquaim | EMBRAPA |
| TRAT-08 | Maranhão | EMBRAPA | TRAT-23 | Yellow oath | EMBRAPA |
| TRAT-09 | Pau Velho | EMBRAPA | TRAT-24 | Miraculous | EMBRAPA |
| TRAT-10 | Surubim I | EMBRAPA | TRAT-25 | Duquinha | EMBRAPA |
| TRAT-11 | Surubim II | EMBRAPA | TRAT-26 | Grandma's taxi | EMBRAPA |
| TRAT-12 | Inajá | UFPA | TRAT-27 | Basti ana | EMBRAPA |
| TRAT-13 | Capembas | UFPA | TRAT-28 | Tomazia | EMBRAPA |
| TRAT-14 | Harpoon | UFPA | TRAT-29 | Purple Eye | UFPA |
| TRAT-15 | Crank | EMBRAPA | TRAT-30 | Santinha | UFPA |

### 8.4 Agronomic characterisations

The varieties were assessed for the following agronomic aspects: average yield, flesh colour, average height, tipping index and rates of attack by the stem borer, white mealybug and root rot.

The average root yield was obtained at 12 months by weighing each plot individually, after cleaning and removing the part that could not be used by the industry. At the same time, on the same plants, the percentage of attack by root rot and the colour of the pulp were assessed.

The average plant height and the percentage of fallen plants were assessed on the 15 plants in the three centre rows of each experimental plot at 10 months of age.

After noting the high rate of infestation by the stem borer and white mealybug in the field, an assessment of the percentage of plants attacked by these pests was carried out per plot on 04/12/2009.

### 8.5 Statistical analysis

Analysis of variance was carried out to check for significant differences between the root yield averages of the treatments (varieties) evaluated. The yield averages were subjected to the Scott-Knott test at a 5% significance level.

**Statistical analyses were carried out using the software -SAEG -** System **for Statistical Analysesll, Version 9.1 (2007), developed by the Federal University of** Viçosa.

# CHAPTER 9

## RESULTS AND DISCUSSIONS

### 9.1 Agronomic characterisations

Table 1 shows the results of the average height, percentage tipping index and flesh colour of the varieties studied. With regard to pulp colour, the greatest importance lies in the diversity of uses it can be put to. In the case of flour production, it is also classified by colour, which can be directly related to the colour of the root pulp. There is also the possibility of other uses, such as the production of tucupi as a utilisation of the manipueira, obtained by pressing the grated manioc, which can add income to the producer. In this case, the use is exclusive to varieties with yellow flesh.

There are regions that prefer yellow-coloured flour and, in the absence of varieties with this natural colouring, flour producers use artificial colourings. Therefore, the availability of high-yielding varieties with a yellow flesh colour, such as the yellow Jurará, Tomázia and Ouro preto varieties, which produced yields of 42.28, 38.38 and 37.29 $t.ha^{-1}$ respectively, is a solution, especially for organic production. In addition, these varieties could also be used to produce tucupi, which is highly consumed in the northern region, especially in the states of Pará, Amazonas and Amapá.

The average plant height varied from 151 cm for the Olho roxo variety to 250 cm for the Surubim I variety.

Plant toppling in the planting area due to wind or other factors, which certainly results in a loss of productivity and root quality, may not be directly related to plant height. Table 1 shows that the varieties with the highest percentage of toppling were Inha, Pacajás, Maranhense, Ararão and Tomázia, with 33.3, 31.1, 28.9, 26.7 and 24.4% of plants toppling, respectively, and their average heights were 194, 203, 204, 222 and 215 centimetres. However, the tallest varieties were Surubim I, Kiriris, Saracura, Surubim II and Bastiana, with heights of 250, 246, 241, 239 and 236 centimetres, respectively, with toppling rates ranging from zero (0.0) to 11.1%.

Table 2 shows the results of the percentage of plants attacked by stem borer, white mealybug and root rot among the varieties studied. With regard to root rot, one of the diseases that causes the most damage in cassava growing, the varieties with the highest percentage of attack were Duquinha, Taxi vovó Paulo Velho and Jurará amarela, with 55.6, 38.9, 33.3 and 27.8 per cent of plants attacked, respectively.

**Table 1:** Average plant height, percentage of fallen plants and root flesh colour of the varieties studied.

| VARIETIES | HEIGHT (cm) | TOMBING (%) | COLOUR OF THE ROOT PULP |
|---|---|---|---|
| Ceará | 204 | 0 | White |
| Saracura | 241 | H,1 | White |
| Pacajás | 203 | 31,1 | Cream |
| Green eye | 230 | 13,3 | White |
| Jurará | 228 | 2,2 | White |
| Inha | 194 | 33,3 | Cream |
| Chico Vara | 194 | 17,8 | White |
| Maranhão | 204 | 28,9 | Cream |
| Old wood | 192 | 20 | White |
| Surubim I | 250 | 4,4 | Yellow |
| Surubim II | 239 | 4,4 | Yellow |
| Inajá | 195 | 15,6 | White |
| Capembas | 190 | 6,7 | Yellow |
| Harpoon | 222 | 26,7 | Yellow |
| Honed | 227 | 4,4 | White |
| BRS Mari | 223 | 0 | Cream |
| Kiriris | 246 | 0 | White |
| BRS Poti | 220 | 2,2 | Cream |
| Paulo Velho | 213 | 20 | White |
| Black gold | 208 | 4,4 | Yellow |
| Yellow oath | 213 | 6,7 | Yellow |
| Miraculous | 190 | 6,7 | White |
| Duquinha | 174 | H,1 | Cream |
| Grandma's taxi | 224 | 6,7 | White |
| Basti ana | 236 | 8,9 | White |
| Tomázia | 215 | 24,4 | Yellow |
| Purple eye | 151 | 0 | Cream |
| Santinha | 199 | 0 | White |

**Table 2:** Percentage of plants attacked by stem borer, white mealybug and root rot.

| VARIETIES | PLANTS WITH | PLANTS WITH | PLANTS WITH |
|---|---|---|---|

| | STEM BORERS (%) | WHITE MEALYBUG (%) | ROOT ROT (%) |
|---|---|---|---|
| Ceará | 1,3 | 6,7 | 5,6 |
| Saracura | 1,3 | 9,3 | 0 |
| Pacajás | 4 | 9,3 | 5,6 |
| Green eye | 2,7 | 0 | 22,2 |
| Jurará | 12 | 1,3 | H,1 |
| Inha | 8 | 0 | 0 |
| Chico Vara | 12 | 4 | H,1 |
| Maranhão | 8 | 0 | 22,2 |
| Old wood | 13,3 | 0 | 0 |
| Surubim I | 8 | 0 | 22,2 |
| Surubim II | 5,3 | 0 | 0 |
| Inajá | 13,3 | 2,7 | 0 |
| Capembas | 8 | 0 | 0 |
| Harpoon | 4 | 0 | 0 |
| Honed | 5,3 | 0 | 22,2 |
| BRS Mari | 1,3 | 1,3 | 5,6 |
| Kiriris | 4 | 0 | 5,6 |
| BRS Poti | 4 | 0 | 0 |
| Paulo Velho | 6,7 | 0 | 33,3 |
| Black gold | 9,3 | 0 | 16,7 |
| Yellow oath | 5,3 | 0 | 27,8 |
| Miraculous | 8 | 0 | 0 |
| Duquinha | 13,3 | 0 | 55,6 |
| Grandma's taxi | 6,7 | 0 | 38,9 |
| Basti ana | 8 | 0 | 16,7 |
| Tomázia | 1,3 | 0 | 5,6 |
| Purple eye | 46,7 | 5,3 | 5,6 |
| Santinha | 17,3 | 0 | 0 |

Extra care should be taken with the introduction of these varieties in the region studied, which until now had not seen any significant records of this disease. However, among these varieties, only the yellow Jurará stood out in terms of root yield, with 42.28 t.ha-1, as can be seen in Table 03. Therefore, as the aim of the research is also to indicate promising varieties for the region, these varieties should automatically be eliminated from the field. Greater attention should also be paid to the Olho Verde, Maranhense Surubim I and Amolada varieties, which showed 22.2 per cent root rot attack (Figure 5a and 5b attached).

The percentage of plants attacked by the white mealybug (Figure 3a and 3b attached) did not exceed 10% in any variety, as can be seen in Table 03. This small outbreak seems to have been a consequence of the severe drought in the region during the period studied, according to data from the INMET Climatological Station in Monte Alegre, PA, which indicated rainfall rates of around 30 mm, 30 mm, 50 mm, 30 mm and 300 mm in the months of August to December 2009, respectively (INMET, 2009). According to FERREIRA (1941), it is common for this insect to appear under these conditions. It's worth noting that with the mandarová control (Figure 6a and 6b attached), carried out on 04/12/2009, a drastic reduction in this insect was also observed and it didn't cause any major damage.

With regard to stem borer attack (Figure 4a and 4b attached), the data in Table 3 showed that all the varieties had some level of attack. However, it's important to note that the varieties Cearense, Saracura, Pacajás, Olho Verde, Ararão, BRS Mari, Kiriris, BRS Poti and Tomázia had incidences in less than 5% of the plants, with Saracura standing out with 1.3%, Kiriris with 4.0 per cent and Tomázia with 1.3 per cent of borer attack which, as well as showing a good level of tolerance to this pest, were also grouped among the best varieties in terms of root yield by the Scott Nott test ($P < 0.05$). 0.05), with 43.18; 41.48 and 38.38 $t.ha^{-1}$, as shown in Table 1.

On the other hand, the varieties Jurará, Chico vara, Pau Velho, Inajá, Duquinha, Olho Roxo and Santinha had the highest incidence of plants attacked by the stem borer, with rates ranging from 12 to 46.7 per cent of plants attacked. In this group, the varieties Olho Roxo and Santinha stood out negatively, with 46.7 and 17.3 per cent, respectively, which had the lowest root yields (Table 1). The low productivity observed in these two varieties was certainly influenced by the high rate of attack by this insect pest.

Among the varieties in the group with the highest root yields, the Inajá variety, with 13.3 per cent, and the Chico Vara variety, with 12.0 per cent borer attack, should be better evaluated, as they were the two with the highest root yields, with more than 45 $t.ha^{-1}$.

As it is a pest that is difficult to control, extra attention should be paid when selecting the manioc seeds, as this is when the insect spreads most. There was a mandarová attack in the experiment and systematic control was carried out by applying Lorban at a dose of 2.5 ml per litre of water.

9.2 **Analysing root production**

The experiment was set up with 30 varieties, as shown in Table 01. The Manivão and Maraquaim varieties were excluded from this total due to their low germination rates, 4.4 per cent and 28.9 per cent respectively.

The results of the average root yields, in $t.ha^{-1}$, of the 28 cassava varieties harvested at 12 months of age were submitted to analysis of variance, the results of which indicated the existence of highly significant differences ($P<0.01$). The coefficient of variation in the experiment was 22.9 per cent.

The average root yields of the varieties were grouped by the Scott-Knott test at 5% significance and the results are shown in Table 3. The average yields ranged from 45.83 to 13.29 $t.ha^{-1}$ for the Inajá and Santinha varieties, respectively. The overall average was 34.91 $t.ha^{-1}$.

Table 3 shows that the varieties Inajá, Chico vara, Inha, Saracura, Jurará amarela, Amolada, Kiriris, Bastiana, Milagrosa, Jurará, Pau Velho, Tomázia, Ouro preto and Taxi vovó were equal to each other and significantly superior to the others, with yields ranging from 45.83 (Inajá) to 35.74 $t.ha^{-1}$ (Taxi vovó). The varieties Olho roxo and Santinha were the worst performers, with 18.33 and 13.29 $t.ha^{-1}$ of roots, respectively.

Considering that the Scott-Knott test works with groupings of means, it is important to note in Table 03 that the Inajá and Chico Vara varieties, despite having statistically equal yields to the Ouro preto and Taxi vovó varieties, produced more than 8.0 $t.ha^{-1}$ than the other two varieties. Based on the results of Polla (2010), who worked with the first two varieties in the same region, this could mean more than 2.0 $t.ha^{-1}$ of flour, which could be economically significant. It is also worth noting that of the 14 best varieties in this experiment, the Inajá, Chico Vara, Inha and Pau Velho varieties were also the most productive in the work carried out by Polla (2010).

**Table 3:** Results of the Scott - Knott test for the root yield averages* ($t.ha^{1}$) obtained from the varieties evaluated.

| VARIETIES | AVERAGE PRODUCTIVITY ($t/ha^{-1}$) |
|---|---|
| Inajá | 45,83 A |
| Chico vara | 45,53 A |
| Inha | 44,43 A |
| Saracura | 43,18 A |
| Yellow oath | 42,28 A |

| | | |
|---|---|---|
| Honed | 41,78 A | |
| Kiriris | 41,48 A | |
| Basti ana | 41,27 A | |
| Miraculous | 39,25 A | |
| Jurará | 38,99 A | |
| Old wood | 38,69 A | |
| Tumazia | 38,38 A | |
| Black gold | 37,29 A | |
| Grandma's taxi | 35,74 A | |
| Old Paul | 34,16 B | |
| Ceará | 33,25 B | |
| Pacajás | 32,81 B | |
| Green eye | 32,7 B | |
| Capembas | 31,89 B | |
| BRS Mari | 31,55B | |
| Maranhão | 31,17B | |
| Surubim II | 30,37 B | |
| Surubim | 29,26 B | |
| Duquinha | 28,94 B | |
| BRS Poti | 28,4 B | |
| Harpoon | 27,29 B | |
| Purple eye | 18,33 C | |
| Santinha | 13,29 C | |

* Averages followed by the same letters do not differ according to the Scott - Knott test at a 5% significance level.

Comparing the results in Table 3 with the average yield for the state of Pará, which is around 16.2 tonnes per hectare[1] (IBGE, 2006), it can be seen that of the 28 varieties studied, only the Olho Roxo and Santinha varieties did not exceed this value. The first five varieties, Inajá, Chico Vara, Inha, Saracura and Jurará Amarela were 183, 181, 174, 167 and 161% more productive than the state average . It should be noted that these results were obtained by liming and fertilising the soil. The variation in average yields between the varieties studied is illustrated in Figure 1.

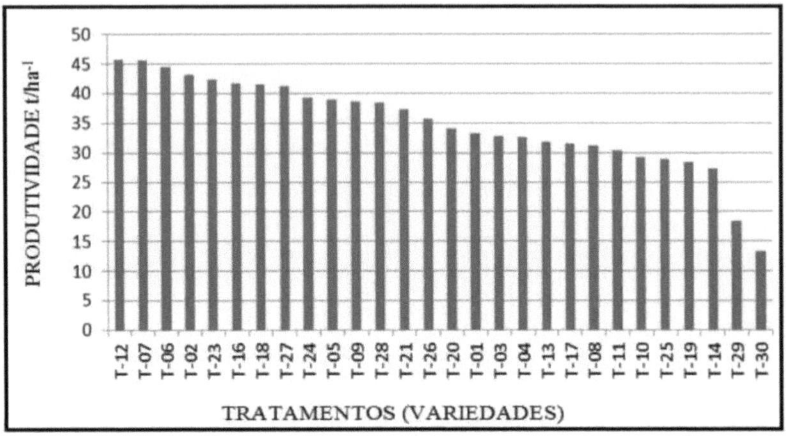


**Figurei:** Average yields of cassava roots from the varieties studied.

Guimarães et al. (2009), evaluating 15 cassava varieties in Vitória da Conquista, in south-west Bahia, obtained root yields ranging from 5.7 to 18.7 $t.ha^{-1}$, well below the results found in this study. Although these are very different varieties and soil and climate conditions, the results highlight the potential of the municipality of Altamira in relation to other traditional regions for this crop, as well as the importance of evaluating available varieties regionally and exchanging genetic material between institutions, because due to the genotype x environment relationship, genetic material that does not respond well in one agrosystem may be interesting in another.

Rodrigues et al. (2007) evaluated 10 varieties of cassava in the municipality of Salvaterra, an agrosystem on the island of Marajó, in the state of Pará. These researchers obtained results ranging from 17.2 to 54.1 $t.ha^{-1}$, i.e. some varieties had higher root yields and others lower than those in this study. However, in addition to working with other varieties and in an ecologically different location, the authors also used organic and chemical fertilisers. This shows that the yield results achieved with the varieties used in this study could also be increased with soil correction and balanced fertilisation, because in both cases the soil is a sandy-textured dystrophic yellow latosol. The varieties Inajá, Chico Vara, Inha, Saracura and Jurará Amarela have comparatively excellent root production potential.

As liming and fertilisation were carried out without the results of the soil analysis, i.e. they were established as a minimum correction because the soil is more than 30 years old and has been used for grazing, the results indicate the need for further studies to calibrate the ideal levels of liming and fertilisation for the local soil and climate conditions.

According to ENGETECNO (2009), the average yield of flour is approximately 25 to 30 per cent, which depends, according to the author, on the variety and the efficiency of the equipment used. Polla (2010), also working with the four varieties with the highest physical yield, in descending order, Inajá, Chico vara, Inha and Saracura, obtained yields of puba-type flour of 28.1, 26.1, 29.3 and 20.8 per cent, respectively.

Based on these results and assuming an average flour yield of 25%, with the yields obtained for these four varieties (Table 03) an average flour yield of around 11.0 $t.ha^{-1}$ can be projected. These results are in line with those obtained by Polla (2010) and demonstrate the high potential of the four varieties for flour production, which allows them to be recommended for the study region.

**10 CONCLUSIONS**

Among the varieties grouped with the highest yields, Inajá, Chico Vara, Inha, Saracura and Jurara Amarela, in descending order, had the highest root yields.

The olho roxo and santinhas varieties stood out negatively with the lowest root yields.

In terms of root rot attack, the most resistant varieties were Pau velho, Surubim II, Inajá, Capembas, Ararão, BRS Poti, Milagrosa, Santinha and the most susceptible were Duquinha, Taxi vovó, Paulo Velho, Jurará amarela, Amolada, Surubim I, Maranhense and Olho verde, all of which had an incidence of root rot greater than 20% .

With regard to the stem borer, the most tolerant varieties were Saracura, Tomázia and Kiriris, with incidences of stem borer attack of less than 5%, while the most susceptible were Olho roxo, Santinha, Duquinha, Inajá and Pau velho, with incidences of attack of more than 10%.

# CHAPTER 10

## BIBLIOGRAPHICAL REFERENCES


ABAM. Brazilian Association of Cassava Starch Producers (Paranavaí, PR). **Dossier on cassava and its derivatives.** Paranavaí, 1998. 34p.

ALMEIDA, F. C. G.; ALMEIDA, F. A. G.; CARVALHO, P. R. **Practical descriptors for botanical characterisation of some cassava cultivars in the state of Ceará.** Fortaleza, 1993.

AREGA D. ALENE, A. D.;NEUENSCHWANDER, P.; MANYONG, V.M.; COULIBALY, O; HANNA, R. The impact of IITA-led biological control of major pests in sub-Saharan African agriculture: **A synthesis of milestones and empirical results.** IITA, Ibadan, Nigeria, 2005. Available at: http://www.iita.org/cms/details/impact/impact_major-pest.pdf, Accessed on 06/06/09.

ARIENTE, M. et al. Competitiveness in the cassava starch industry: an exploratory study. **Revista da FAE**, Curitiba, v. 8, n. 2, p. 53-60, 2005.

BELLOTTI, A. C.; SMITH, L.; LAPOINTE, S. L. Recent advances in cassava pest management. **Annual Review of Entomology**. v. 44, p. 343-370. 1999.

BELLOTTI, A.C. **Arthropod pests.** In: Cassava: Biology, production and utilisation. Eds: Hillocks, R. J., Thresh, J.M., Bellotti, A. C. 2002. CAB International. Oxon, UK. 332 p.

BELLOTTI, A.C. ; ARIAS, B.V. ;VARGAS, O.H. ; REYES, J.A.Q. ; GUERRERO, J.M. **Insects and mites harmful to yuca and their control.** In: The yuca in the third millennium: modern systems of production, processing, utilisation and commercialisation. OSPINA, B.; CEBALLOS, H. (eds). CIAT/CLAYUCA, n.327. 2002b. 586 p.

BELLOTTI, A.C. ; ARIAS, B.V. ;VARGAS, O.H. ; PENA, J. E.. **Losses in yuca crop yield caused by insects and mites. In: Yuca in the third millennium: modern systems of production, processing, utilisation and commercialisation.** OSPINA, B. ; CEBALLOS, H. (eds). CIAT/CLAYUCA, n.327. 2002c. 586 p.

BOBBIO, P. A.; BOBBIO, F. O. **Chemistry of food processing.** São Paulo: Varela, 1992.

BORGES, M. de F.; CARVALHO, V. D. de; FUKUDA, W. M. G. Effect of heat treatment on the post-harvest conservation of table cassava (Manihot esculentaCrantz) roots. **Revista Brasileira de Mandioca,** Cruz das Almas, v. 11, n. 1, p. 7-18, 1992.

BOURSIER, B. Applications alimentaires des amidons modifiés. **Industries Alimentaires et Agricoles,** Paris, v. 111, n. 9, p. 583 592, 1994.

CAGNON, J. R.; CEREDA, M.; PANTAROTTO, S. In Cd-rom. **Series: Culture of Latin**

**American starchy tubers.** Vol.2 - Cultivation of Latin American starchy tubers. Cargill Foundation. Aug/2002.

CÂMARA, G. M. S.; OLIVEIRA, E. A. M.; RUIZ, S. T.; PUZZO, R. B.; FURLAN, N. M.;
MAULE, R. F. **Quantification of branch losses and their effect on root production.** Revista Brasileira de Mandioca, Cruz das Almas, v. 13, n.2, p. 100-109, 1994.

CARDOSO, E. M. R. **Products and by-products.** In: SOUZA. Embrapa manioc processing and utilisation. Brasília. Embrapa manioc and tropical fruit growing, 2005. pag.22.

CARVALHO V. D. **Cyanogenic glycosides.** In: TRAINING COURSE **FOR TECHNICIANS** IN **THE MANDICA CULTURE, 12, 1978, Lavras. Anais...Lavras, 1978.** 24 p.

CARVALHO, C. R. L. et al. **Chemical analyses of foods.** Campinas: ITAL, 2000.

CEREDA, M. P. **Valorisation of by-products as a way of reducing production costs.** In: CEREDA, M. P. (Coord.) Manejo, uso e tratamento de subprodutos da industrialização da mandioca. São Paulo: Cargill Foundation, 2001. V. 4, pag. 305 (Latin American Tuber Crops Series).

CEREDA, M. P. et al. **General properties of starch**: volume 1. São Paulo: Cargill Foundation, 2001.

CEREDA, M. P. **Tradition of manioc flour production in Amazonia.** In: SOUZA. Embrapa manioc processing and utilisation. Brasília. Embrapa manioc and tropical fruit growing, 2005. pag.145.

**CURCELLI, F. et al. Cyanogenic content of cassava aerial part "in natura" and in the** form of hay. **Revista Raízes e Amidos Tropicais,** p. 1185-1189, 2009.

CEPLAC. **Cassava, 2007.** Available at. <www.ceplac.gov.br/radar/mandioca>Accessed on 16 January 2010.

CHAN, S. K., MOHD ISA, H. **Effects of planting methods and other cultural factors on the yield of cassava.** Teknologi Pelbagai Tanaman, v. 1, n. 1, p. 13-20, 1985.

CONCEIÇÃO, A. J. **Cassava.** São Paulo: Nobel, 1983. 382 p.

CORREA, A. D.; FARIAS, A. R. N.; MATTOS, P. L. P.. **Utilisation of cassava and its products in human nutrition.** In: SOUZA. Embrapa manioc processing and utilisation. Brasília. Embrapa manioc and tropical fruit growing, 2005. pag.224.

CORRÊA, H., VIEIRA NETO, J. C. **Agronomic aspects of cassava cultivation.** In: CORRÊA, H., CALDEIRA, A. (Ed.). Intensive course to train technicians in cassava cultivation. Lavras: ESAL, 1978. p. 314-364.

CORREA, H.; ROCHA, B.V. **Manejo da cultura da mandioca**. Informe Agropecuário, v.5, n.59/60, pag.16-30, 1979.

DINIZ, M. de S.; OLIVEIRA, A. M. G.; COUTINHO, S. da C.; SANTANA M. do A. **Behaviour of Cassava Varieties in the Municipalities of Porto Seguro and Santa Cruz Cabrália-BA**. Research and Development Bulletin; EMBRAPA, Cruz das Almas,

Bahia, 2004. 16 p. Available at<www.fitoterapia.com.br/portal >. 2009. Accessed on: 16 January 2011.

EMBRAPA. National Soil Research Centre. **Brazilian soil classification system**. Brasília: Prod. De informação; Rio de Janeiro: EMBRAPA Solos, 1999.

EMBRAPA MANDIOCA E FRUTICULTURA. (2003_ Available at: <http://www.sistemadepruducao.cnptia.embrapa.brfontesHTML/Mandioca/mandioca _centros ul/importância. http>. Accessed on 01 January 2017.

ENGETECNO. **Manufacture of cassava flour**. 2009. Available at:< http://www.engetecno.com.br/port/tecnologia_farinha_de_mandioca.htm >. Accessed on: 26 May 2011.

FAOSTAT database. Available at:< http://faostat.fao.org/site/567/default.aspx>. Accessed on: 16 January 2011.

FAO - **Food and Agriculture Organisation of the United Nations**. Available at:< http://faostat.fao.org>. 2007. Accessed on: 16 January 2011.

FERREIRA, LIMA, A. D 1941 - **The yucca weevil.** Secr. Viaç. Obr. Publ. Agric. Est. Santa Catarina, Bol. 2, 6p., 4 ests.

FIALHO, J.F.; OLIVEIRA, M.A.S.; ALVES, R.T.;PEREIRA, A.V.; JUNQUEIRA, N.T.V. & GOMES, A.C. **Damage of the Lace Bug on Cassava Productivity in the Federal District**. Technical Communication - Embrapa Cerrados, Planaltina, DF. n. 48. p.1-3. 2001.

FIALHO, J. DE F.; VIEIRA, E.A.; PAULA-MORAES, S.V.P.; SILVA, M.S.; JUNQUEIRA, N.T.V. Damage caused by lacebug in the production of aerial part and roots of cassava. **Scientia Agraria,** Curitiba, v.10, n.2, p.151-155, 2009.

FOLEGATTI, M. I. da S.; MATSUURA, F. C. A. U.; FILHO, J. R. F. **The cassava flour industry**. In: SOUZA. Embrapa manioc processing and utilisation. Brasília. Embrapa manioc and tropical fruit growing, 2005. pag. 63.

FREITAS R. A. et al. A rheological description of mixtures of a galactoxyloglucan with high amylose and waxy corn starches. **Carbohydrate Polymers,** Barking, v. 51, p.

25-32, 2003.

FROESCHNER, R. C. The neotropical lace bugs of the genus Vatiga (Heteroptera: Tingidae), pests of cassava: new synonymies and key to species. **Proc. Entomol.** Soc. Whash. n.95, p.457-462. 1993.

FUKUDA, W. M. G.; BORGES, M. de F. Influence of harvest age on root quality in different cultivars of table cassava. **Revista Brasileira de Mandioca**, Cruz das Almas, v. 9, n. 1-2, p. 7-19, 1990.

FUKUDA, W. M. G.; SILVA, S. O. **Cassava improvement in Brazil** . In: Latin American Tuberous Starchy Crops. Cargill Foundation, v. 2, p.242-255, 2003.

GIRAUD, E.; BRAUMAM, A.; KELEKE, S.; GOSSELIN, L.;RAUMBAULT, M. Contrôle de la fermentation du maiúoc pour un meiller gari: utilization d' un starter de lactobacillus plantarum à activité linamarase et amylase. In : FOLEGATTI, M. I. da S.; MATSUURA, F.

C. A. U.; FILHO, J. R. F. **The cassava flour industry**. In: SOUZA. Embrapa manioc processing and utilisation. Brasília. Embrapa manioc and tropical fruit growing, 2005. pag.66.

GUIMARÃES, D. G.; VIANA, A. E. S.; MUNIZ W. F.; CARDOSO, A. D.; GOMES, I.R.; FERNANDES, E. T.; MAGALHÃES, G. C.; ANJOS, D. N. dos; FOGAÇA, J. J. N. L.
**Evaluation of cassava varieties in Vitória da Conquista - BA.** XIII Brazilian Cassava Congress, 2009

IBGE. **Municipal Agricultural Production** 2009. Rio de Janeiro: IBGE, 2009. Available at.< http://www.ibge.gov.br>. Accessed on: 02 March 2017.

IBGE. Brazilian Institute of Geography and Statistics. Systematic Survey of Agricultural Production - Cassava Crop 2009/2010. 2010

IBGE, **Municipal Agricultural Production** 2013. Rio de Janeiro. Available at:<http://www.sidra.ibge.gov.br>. Accessed on: 16 January 2017.

INMET. **Maps from the Agroclimatological Bulletin**. INMET, 2011. Available at< http://www.inmet.gov.br/climatologia/bol_agro/mapas.php.

LORENZI, J. O. Variation in the culinary quality of cassava roots. **Bragantia**, Campinas, v. 53, n. 2, p. 237-245, 1994.

LORENZI, J. O. et al. Evaluation of table cassava cultivars in the Ribeira Valley (SP). **Bragantia**, Campinas, v. 55, n. 1, p. 141-146, 1996a.

LOZANO, J. C., BOOTH, R. H. **Diseases of the yuca.** In: CENTRO INTERNACIONAL DE AGRICULTURA TROPICAL. Yuca, research, production and utilisation. Cali,

1982. p. 421-462.

MARRIOT, J.; BEEN, B. O.; PERKINS, C. The aetiology of vascular discolation in cassava roots after harvesting: association with loss from wounds. **Physiologia Palntarum**, Copenhagen, v. 44, n. 1, p. 58-42, 1978.

MARTINAZZO, M.; KRAEMER, B.; CASTOLDI, G.; FIORESE, S.; LOHMANN, T.R.; PIETROWSKI, V. **Influence of the lace bug on cassava productivity in the western region of Paraná**. In: XII CONGRESSO BRASILEIRO DE MANDIOCA, 2007. Paranavaí. CERAT, Botucatu. 2007b.

MATOS, P. L. P; CARDOSO, E. M. R. **Cultivation of Cassava for the State of Pará**. Available on the website: http://sistemasdeproducao.cnptia.embrapa.br/FontesHTML/Mandioca/mandioca_para/index.h accessed on 21 January 2011.

MATTOS, P. L. P. **Technological development for cassava cultivation**. Cruz das Almas: EMBRAPA/CNPMF, 1993. 45 p. (Documents, EMBRAPA-CNPMF, 51).

MIDIO, A. F. & MARTINS, D. I. **Toxicologia de alimentos.** Ed. Varela. São Paulo, 2000.

NEUENSCHWANDER, P. Biological control of the cassava mealybug in Africa: a review. **Biological Control.** v. 21, p - 214-229, 2001.

OKA, M.; LIMSILA, J.; SUPACHAI, S. Relationship between characteristics and germination ability of cuttings in cassava *(Manihot esculenta* Crantz). **Japan Agricultural Research Quarterly**, Tokyo, v. 21, p. 70-75, 1987.

OLIVEIRA, C.M.; FIALHO, J.F. **Biology of the cassava root mealybug Prototonia nevesi Fonseca** (Hemiptera:Margarodidae) in the laboratory. In: XI CONGRESSO BRASILEIRO DE MANDIOCA, 2005, Campo Grande. Abstracts... Campo Grande: EMBRAPA, 2005.

POLLA, C. **Evaluation of cassava varieties in the Princesa do Xingu community, Altamira - PA**. Final course work, Altamira-PA, UFPA, 2010.

PADONOU, W; MESTRES, C.; NAGO, M.C. The quality of boiled cassava roots: instrumental and relationship with physicochemical properties and sensorial properties. **Food Chemistry**, v.89, p.261-270, 2005.

PEREIRA, A. S.; LORENZI, J. O.; VALLE, T. L. Evaluation of cooking time and cooked dough pattern in table cassava. **Revista Brasileira de Mandioca**, Cruz das Almas, v. 4, n. 1, p. 27-32, 1985.

RICKARD, J. E. Deterioration of harvested yuca roots. **Yuca Boletin Informativo**, Cali, v. 8, n. 2, p. 3, Nov. 1984.

RICKARD, J. E. Physiological deterioration of cassava roots. **Journal of the Sciense of Food and Agriculture**, London, v. 36, n. 3, p. 167-176, Mar. 1985.

RODRIGUES, J. E. L. F.; TEIXEIRA, R, N, G.; BOTELHO, S. M. **Competition between cassava varieties (*Manihot esculenta* L.) for family farming in the municipality of Salvaterra, PA.** Technical communiqué 189. Belém, PA, 2007.

SAGRILO, E.; OTSUBO, A.A.; SILVA, A. de S. Productive performance of cassava genotypes in the Ivinhema Valley, MS. **Revista Raizes e Amidos Tropicais**, Botucatu, v.3, Oct. 2007. Available in: http:/ .www.cerat.unesp.br/revistarat/volume3/153%20Edvaldo%20Sagrilo.pdf In: 18/04/2008.

SANTOS, J, C. **Development of the cassava crop as a function of the position of the manioc when planting.** Final course work, Altamira, UFPA, 2007.

SANTOS, M.F.; PAULA-MORAES, S.V.; VIEIRA, E.A.; FIALHO, J.F. OLVEIRA, C.M.;TAKADA, S.C.S. SOUZA, A.A.C. **Hydrocyanic acid (HCN) content as a parameter for selecting possible cassava accessions resistant to the lace bug.** In: II SIMPOSIO INTERNACIONAL SAVANAS TROPICAIS and IX SIMPOSIO NACIONAL CERRADO, 2008, Brasília. Proceedings... Embrapa. Brasília, DF. 2008a.

SILVA, C. S. e; AUGUSTO, S. G.; ANDRADE, A. U. **Agrometeorological characterisation of Monte Alegre, PA.** In: Proceedings of the agrarian sciences integration week. Altamira, PA, 2015.

SOUSA, L. da S.; FIALHO, J. F. **Cassava root scrapings. 2007.** Available at: www.sistemaprodução.cnptia.embrapa.br/fonteshtml/mandioca/mandioca_cerrado/pr ocessam ento.htm. Accessed on 4th March 2011.

SOUZA, A. S. **Selection and preparation of planting material.** In: EMBRAPA. Practical instructions for growing cassava. Cruz das Almas: EMBRAPA /CNPMF, 1993. p. 42-52.

TONTINI, F. **Avaliação da densidade populacional e posição da maniva na produtividade de mandioca, Altamira-Pará.** Final course work, Altamira, UFPA, 2009.

VALDIVIÉ, et al. **The cassava root.** In: VALDIVIÉ, M. I. N.; BICUDO, S. J. (Eds). Feeding monogastric animals: cassava and other non-conventional foods. Botucatu, Brazil. FEPAF, 2011. p. 1-18

VIANA, A. E. S.; SEDIYAMA, T.; LOPES, S. C.; SEDIYAMA, C. S.; ROCHA, V. S. Effects of length in stem cutting and its planting position on cassava yield. **Acta Scientiarum**, Maringá, v. 22, n. 4, p. 1011-1015, 2000.

WANAPAT, M. **Potential uses of cassava as ruminant feeds.** In: The use of cassava root and leaves for on-farm animal feeding, Bangkok: CIAT, 2009, 122p.

ANNEXES

**Figure 2.a.** Selection of seed maniocs for planting
**Figure 2.b.** Soil correction with application of dolomitic limestone

**Figure 3.a.** Symptom of a plant attacked by the white mealybug
**Figure 3.b.** Symptom of a plant attacked by the white mealybug

**Figure 4.a.** Stem borer attack in pupal stage
**Figure 4.b.** Stem borer attack in larval form

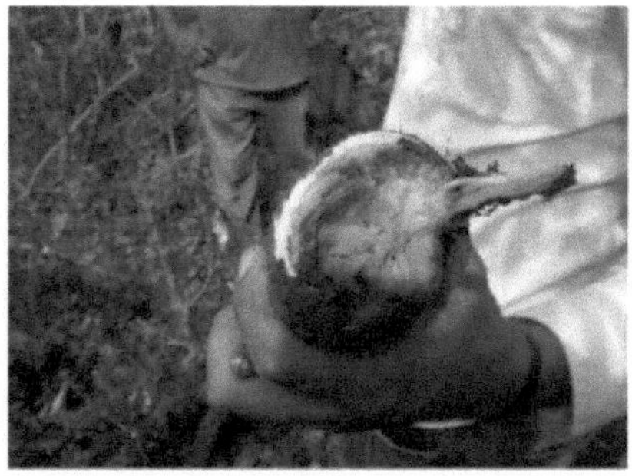

**Figure 5.a.** Plant with root rot symptoms
**Figure 5.b.** Plant with root rot

**Figure 6.a.** Mandarovah presence in the experiment
**Figure 6.b.** Traces of the presence of Mandarová

MIX
Papier aus verantwortungsvollen Quellen
Paper from responsible sources
FSC® C105338

Printed by Books on Demand GmbH, Norderstedt / Germany